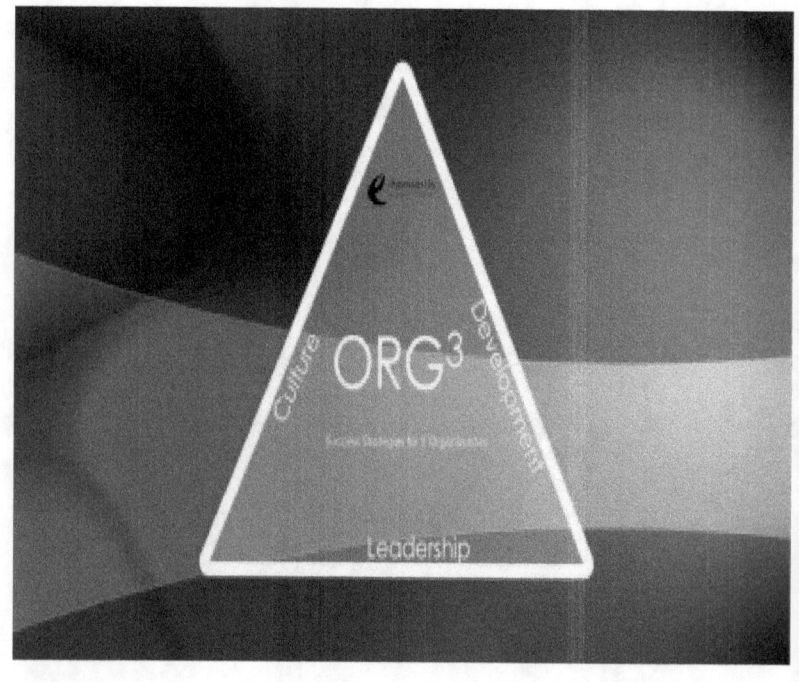

Empower Us, Inc. All Rights Reserved

ORG 3

Culture. Development. Leadership

Series Introduction

Dr. Melanie Magruder

Empower Us, Inc. Atlanta, GA,

Table of Contents

- Foreword v
- Organizational Culture 7
  - What is Culture? 7
  - Cultural Climate 7
  - The New Regime 8
  - Cultural Socialization 9
  - Customer Loyalty 10
  - Hiring Practices 11
  - Darren's Move 12
- Organizational Development 16
  - Starting Out 16
  - Planning for Success 18
  - In the Interim 19
- Organizational Leadership 20
  - The Foundation 20
  - Queue on Clues 20
  - No Gray Areas in Ethics 24
  - Positive Influences 26
- Closing 28

Foreword

Org3

The need is growing for information that is accessible to all members of the IT Organization to empower and inform them on how to proceed in their profession. Too often individuals are hired into positions without any former knowledge about how they should take up their new roles. Employees are hired or promoted into management positions with no prior leadership experience. They may not be right for the position; rather, someone who makes decisions merely believes they are. People who may be entering the workforce for the first time, straight out of college, have no idea what they are about to face daily. Most likely, these graduates are interested in obtaining a position to pay their bills, take care of family members, and pay student loans. These are some of the primary reasons people seek employment; however, many also want to be fulfilled and enjoy what they do each day. They prefer to think of their jobs not only as work but also to provide a service based on their abilities, skills, and talents. Whatever the reason, an employee in any organization, with the proper tools, will be prepared for what comes next.

Empower Us, Inc. has created the Org3 series to promote success in an IT organization's culture, development, and leadership. This first installment in a projected seven-book series provides an overview of key topics that will be presented in the series. Though Org3 focuses specifically on IT organizations, the principles presented can be applied to most, if not all, organization in any industry. In the series, practical application, examples, and real stories are included to demonstrate the reality of the successes and failures and what can change.

Organizational Culture

"People might not get all they work for in this world, but they must certainly work for all they get." Frederick Douglass

What is Culture?

The culture in an organization includes such aspects as how things are done, what the members can wear, and what methods are used for communication and interaction. The culture of the organization is the foundation that generates any strategic outcomes. Culture is determined by leadership because leaders' behaviors are followed by other members of the organization. It is important for leaders to create a culture that nurtures and provides growth to all members of the organization. It should be a place where employees want to come because they enjoy it, not just because they must work to earn a living. The more people enjoy where they work and with whom they are working, the more they will invest in the success of the organization.

Cultural Climate

The cultural climate within an organization affects the productivity, and hence the profits and losses of the organization. An organization succeeds when all members share a common goal, vision, and cooperation. The organization that practices integrity and honesty as a code of ethics tends to lead employees in a positive direction. A hierarchical leadership model can result in a kind of domination that sets a tone in which employees feel coerced, through threats or force, to make decisions that are not compatible with their own ethical standards. Leaders are empowered to direct the organizational culture, negatively or positively.

The New Regime

>Marcia had worked at XYZ Corporation for over 10 years under two CEOs and several managers. Leadership turnover was frequent; employees could not keep up with the changes. New CEOs always brought in teams with whom they were familiar and with whom they had worked in the past. Because the direction of the organization was not stable, it was difficult to carry out the objectives. It was not possible to determine the organization's vision or mission. One CEO ran the company like a military operation, whereas the other treated it like a democracy wherein everyone had a voice. The climate was very shaky.

What would you do?

In such situations, it is important for leaders to set the tone. Whether there are a few or many hierarchies, someone in a leadership role must take the reins, even if it is only for one department, and establish a more settled, secure workplace.

Cultural Climate

1. The tone is set by leadership
2. In Maslow's Hierarchy of Needs, the first tier (physical needs) includes monetary provision and the second, safety, relates to how an employee is treated and the stability of the organization. Thus, at a minimum the first two tiers need to be met at one's place of employment.
3. Constant turnover makes for a very unstable cultural climate, because if the employees do not believe they have stable employment, it will be difficult (almost impossible) for many to believe they should not seek employment elsewhere.

Cultural Socialization

Organizations are often not as socialized as they should be in terms of their cultural values. If organizational members do not know what they are—which is often the case—it is difficult for them to embrace or exercise such values. New organizational members may be made aware of these values through orientation; nevertheless, to incorporate them into workplace activities, a demonstration of the values is necessary in terms of what the company is built around and how members are expected to interact with customers and each other. Newcomers to the organization are often more vigilant about incorporating these values into their daily interactions within the workplace than the members who have been there for a while. The cultural values message must be socialized and demonstrated to organizational members. People believe what they see and will ultimately follow those examples rather than presentations.

As organizations experience the challenges of change and diversity, it is more relevant than ever to understand the influence of culture. Cultures vary in what they value in terms of an individual's contributions to work.

Customer Loyalty

> *Mike has been a sales manager for over 20 years. Worked for the same company, hardly takes a vacation, and believes in the organization's philosophy:* **"Customer Loyalty is what keeps the doors open."** *Recently, he has noticed a change in how long-standing customers are being treated. His manager has been with the company for approximately one year; the previous manager lasted nine months. Mike is not interested in the management position; he likes having a work-life-balance and knows that that job will interfere with his belief.*
>
> *One of Mike's customers contacted him with some very upsetting information, which he found hard to believe. The*

customer told Mike that the new manager had refused to honor a rebate that had been offered him for many years.

Although the promise for the rebate is not in writing, this customer generates over $2 million in revenue for the company. What could possibly be the reason for this, Mike thought? After all, **Customer Loyalty is what keeps the doors open***, right?* Mike brought this concern to his manager's attention. He was told that because the rebate offer is not in writing then he could no longer honor it, even if it meant risking the customer's business.

What would you do?

The new manager failed to see that he was going against what the company had socialized regarding customer loyalty. Verbal contracts should be as binding as those that are written. Be mindful of what is being demonstrated v what is being said. It matters.

1. Do what you tell others to do.
2. Be clear about goals and communicate.

Hiring Practices

Organizations' recruiting policies are formulated with awareness of the messages they send about company values. Organizational members value fairness, honesty, achievement opportunities, and attention to their concerns. Pay is also high on this list. How members are brought into the organization is a key component in demonstrating the implementation of values. Hiring managers may want to bring people onboard whom they know because they may think it will be easier to work with people they know than with someone whom they have never met. They also may take advice from other organizational members if they believe a certain individual is the best candidate for the position. Hiring managers have been known to call their friends and offer them a position before the interview takes place, in which case the interview becomes only a formality. Friends are hired and promoted, even if they are not qualified for the role. Members of the organization may also pass over qualified employees for positions that would be entirely suited to them because of bias or some other personal judgment. The fact that some organizational members place their personal beliefs over doing what is best for the organization's strategic goals results in a need for diversity programs. If companies were fair and equal in their culture, they would not need a diversity program.

Darren's Move

> An executive level recruiting company contacted Darren Morgan, a seasoned IT executive and current CIO at one of the top 10 Technology firms in the U.S. He had been at the company for a little over 5 years, and at other major firms

prior to that. Darren was well known in the tech community, and he had strong relationships both in and out of the company where he was currently employed.

Darren was committed to the success of the company and, through leadership methods, ensured that his subordinates were equally committed. He interviewed for the new position. When his current employer found out, he decided to walk him out before he could give notice. (When proprietary information is at stake, a resource is asked to leave the company for obvious reasons). Ten minutes later, Darren scrolled through his phone and began calling his fellow employees in the company. He explained his new opportunity to them and asked them to leave the company to work with him. He guaranteed employment for them and anyone else who wanted to join them at the new organization. Each person who decided to join Darren was hired without a formal interview. Where did these actions leave their previous employer? What happened to the loyalty that had been so adamantly conveyed by those who left for a shiny new object?

What would you do?

Most people are loyal to their personal goals, not necessarily those of the company in which they work. Hiring is not always fair, although it should be.

1. Hire candidates who are qualified for the position.
2. Demonstrate fair hiring practices, and don't just write them down. Implement them!
3. Be forthcoming regarding your motives.

Embracing Multi-Generational Culture

As diversity in the workplace continues to evolve, each organizational member must learn how to embrace it in a positive manner. The generational focus on Baby Boomers (1946-1964), Gen Xers (1965-1981), and Millennials (1982 – 2002) is clear and apparent simply because these groups are all in the workplace now. Thinking of these groups as a team within the organization is the best approach. Why? Mainly because each member has a unique talent or skillset to contribute that will mean success for all. People are quick to segregate into groups based on many various factors; and, in turn, groups adopt a segregation mentality in lieu of embracing differences. With each generation, there are differences that stand out and that are measured by the preceding one.

Some Baby Boomers were living a "free" lifestyle in the 1960s. They were criticized for being radical, drug users, and lazy, amongst other things. Over the past 30 plus years, many have become executive leaders and entrepreneurs in many organizations (large and small); hence, they can offer their experience and expertise to others. They can pass on their knowledge, thus keeping the torch lit for those who will soon stand in their place.

Gen X was labeled as the first generation in centuries to have minimal supervision and parental guidance because of increasing divorce rates and both parents' being in the workplace. This group is filled with successful entrepreneurs and organizational executive leadership. They, too, are available for sharing wisdom on business success and can be a catalyst for organizational growth opportunities.

Currently, there are the Millennials, labeled as the "NOW" generation. They are very savvy with social media and technology, and are very liberal in their approach to life. This group can provide a new perspective on real-time communications and a modern perspective on entrepreneurship.

If one looks at the various categories into which these groups have been placed, one would never think they would have turned out the way they have. Just think about how people view differences more than similarities. When there is a T.E.A.M. (Together Everyone Achieves More) approach to collaboration in the workplace, the differences are what make the effort a success.

Embrace Change

If everyone would simply embrace what everyone can bring to the organizational table, the successes would be so much greater than one could imagine. If everyone would stop judging and placing each other in a labeled box based on a birth year, collaborative efforts would rule! Stop the generational segregation cycle and embrace the multi-generational culture because we can all learn from each other.

1. Be open minded.
2. Invest in engaging with those who are different; seize the opportunity.
3. Don't judge based on a stereotype.

Organizational Development

"We may encounter many defeats but we must not be defeated."
Maya Angelou

Starting Out

Information on the development of organizations can be directed at both startups and established companies. It is obvious why a startup may need supplemental information to support stabilizing and building a strong organization. When new organizations or entities are created, the development is organic and informal at first. Because the startup is new, there is a need to establish the organizational strategy and to align procedures and processes to meet them; hence formalization, specialization, standardization, complexity, and mechanics all follow in time. But what about established organizations that have been in business for a long time?

Sometimes the cobwebs need to be dusted off and innovative ideas inserted into moldy old processes that worked well many years ago. The times, they are changin'.

Employees are the foundation of any organization; thus, it cannot be developed without them. Organizational leaders must understand that there are three types of employees: (1) those who love what they do and don't care where they do it, (2) those who are loyal to their leadership, and (3) those who believe in the success of the company where they work. An organization's strategy may also be a crucial factor when evaluating employee commitment or intentions to stay and in helping the organization's employee development activities.

Organizational development requires an ability to think about situations in more than one way. This is a complex process that requires quality judgment based on the information on hand. The process must be innovative to keep up with the current trends of what would work best today, not 20 years ago. This takes effort, time, practice, and feedback. It is necessary to engage those who will actually be responsible for assuring the strategies are carried out. Working in silos or with a select few is a sure way to fail at bringing about the best possible outcomes.

Planning for Success

An organization without a plan can be labeled as reactive, shortsighted, and rudderless. It is essential to plan and follow up to assure that the goals and objectives are met in developing the organization. However, because the organizational development process is fluid, it should be revisited periodically to confirm that the organization is on track and to see whether adjustments need to be made.

Changing organizational processes requires innovation and a clear understanding of existing strengths and culture. IT organizations are challenged with ever-changing and innovative technology. The new hot code last year has likely been replaced with something different. The IT Organization may need to adjust in their organizational strategies and development practices more than other organizational industries do.

Another building block for organizational development is to promote member self-worth to the organization. Their self-worth within the organization reflects how the members are treated, what they see others do, and how they feel about themselves in their role. Members who are encouraged to reach their potential contribute to the success of the organization and leader. Conversely, leaders who are superficial, egocentric, insincere, and manipulative can lower an organizational member's self-worth.

In the Interim

Karen is an accountant in a mid-size CPA firm. She is the go-to person for problem solving and locating discrepancies. Karen is a CPA with an MBA from one of the top Ivy League schools in the U.S. She loves her job, and she has shown leadership qualities on many occasions, has been a model team builder, and has trained three of the new managers. Karen has applied to be manager each time the position has opened, but each time, she has been told she is not ready. A search is being conducted for a new manager, and the COO has asked Karen to be the interim manager of the department until someone is hired. Why would they ask her to act as manager and not promote her? Even though Karen loves her job and does it very well, she knows it is time to look for employment at an organization that will appreciate her talents and will demonstrate it by way of compensation and job title.

What would you do?

Building self-worth starts with acknowledging employees' talents and rewarding them. If leaders make biased decisions, employees feel unappreciated and unsuccessful.

1. Reward employees for their contributions.
2. Build their sense of self-worth by employing their talents.

3. Be fair in promoting those who are qualified.

Organizational Leadership

"Management is doing things right; leadership is doing the right things." Peter Drucker

The Foundation

The most crucial factor to be presented to organizations is leadership, which is the foundation of any organization. Effective leadership is contingent on the specific style leaders use to influence people's behavior so that they will make an effort to change. Leaders may increase the confidence of their followers; thus, elevating followers' need for and expectations of success.

Queue on Clues

Cluelessness sometimes possesses those whom one would think it should not. The brightest managers may maneuver or meander their way to the apex of enterprises great and small. Then they do dumb things. Sometimes they may be too smart for their own good. Such may be enhanced by pride and arrogance, or even by an unconscious desire to fail. The best and brightest people can have as many psychological problems as the intellectually challenged. The primary source is not personality or IQ; rather, it comes from peoples' inability to have a broad view of whether they are on track and the willingness to make adjustments if they are not.

Leaders stand at the edge of different fields from most people because they stand between dreams and reality. They have to attract the energy to make the dreams last, and at the same time they have to cope with reality, which brings unexpected events. Leaders are judged on results, not intentions. Leadership training gives leaders the skill to implement a plan and the proficiency to put the plan into action. Well-informed leaders perfect the art of decision-making and promote innovation. Leadership development creates new awareness of various approaches that can be used in the organization.

Good leaders are efficient, effective, and precise in the way they incorporate and dispense information throughout the organization. Though experts may disagree over the exact leadership methodology that should be used, the leadership method chosen should be a good fit and should make a substantial difference in the organization.

A good leader understands this and creates an atmosphere within the organization that empowers, inspires, builds, and leads employees to be more proactive in their roles. Leaders must be open to participation from the followers in their efforts to promote an environment that gives employees the freedom needed to become better leaders. The core values of the organization, as implemented through leadership, promote growth for both employees and the organization. Leadership goes beyond ordinary expectations by communicating a sense of mission, stimulating learning experiences, and inspiring innovative ways of thinking.

A strong leader operates through experience, not education. They have achieved goals by using various motivational profiles, operating and adhering to differing work assumptions, and pursuing differing incentives. Leaders often fail to understand the power of culture in organizations to influence values and behaviors. The challenges, hardships, successes, and performance are major influences on the leader's ability to direct an organization strategically.

Leaders also act as change agents within organizations, as they are responsible for directing, organizing, and facilitating change. It takes a unique individual to be a change agent who possesses special characteristics. Some key characteristics of change agents are creativity, courage, visibility, perseverance, and the driving motivation to make the change occur. When leaders act as change agents, they are not only leading the change but also actively participating in the change effort.

The leader must clarify specific values and communicate what these values mean not only to the organization but also to individuals within the organization. It is important to align the values of an organization with the values of its members. Thus, organizational leaders face challenges the other members may not encounter—for example, acknowledging the unethical practices of their peers or firing an employee for breaching the standards of the organization's culture.

Leading with Integrity

Leadership is the foundation of not only great, but also all, organizations. Doing the right thing for the organization entails character, integrity, and self-discipline. An organization that supports moral integrity and ethical decision making should be promoted. Employees who believe their leaders condone unethical behavior are inclined to behave unethically as well. The strategies implemented by leadership need to be planned carefully to prevent negative consequences.

Leaders may create a negative impact on the organization by imposing increasingly complex and demanding deadlines. Yes, demands and deadlines can make great organizations; however, it is not *what* is done but *how* it is done that makes an enormous difference.

A productive leader generates successful outcomes by demonstrating positive examples that are honest and forthright. Enabling organizational members to innovate prepare for strategic changes, and address challenges to sustain high performance standards are a few qualities of effective leaders, who hone integrity at their core. Building relationships and team cohesiveness are also important talents for leaders to possess.

The end mission of a leader is to help individuals and organizations reach the essential goal. Leaders must decide how this will be accomplished. What style of leadership are you demonstrating? From the several styles available, I am partial to transformational leadership because it focuses mainly on the growth of the follower and includes implementing guidance to meet or exceed organizational goals. What is the difference between managing and leading? In a nutshell, managers focus more on the task to be accomplished; the leader focuses on the person who is performing the task. Have you been trained to lead or are you focused on the process or task at hand? Some who are hired or promoted to a leadership role have no experience. They often fail at meeting the "essential goal" simply because their leadership skills have not been developed.

No Gray Areas in Ethics

Leaders can prevent organizational failure by building ethical systems with proper procedures that encourage members to use respectful, committed methods for speaking out on ethical issues. Integrity and honesty can also result in adverse effects from a human resources point of view if it is understood that certain shared interests could result in negative consequences for the organization.

The ethical climate of an organization has a direct effect on trust in working relationships. "Ethical climate" refers to the members' perceptions of the ethical standards reflected in the organization's practices and procedures. It is important for leaders to promote a positive ethical climate, as it will create positive employees. Too often, communication issues arise owing to closed door meetings and dissemination of information to subordinates through an organization's hierarchy. Employees who believe they work for an organization that cares about them individually will prove to be more committed to the goals, mission, and vision of the organization.

Leaders who are more concerned with their personal success than with the success of the organization are prone to promote organizational gossip, suspicion, and communication failures. Organizational members who experience communication issues question their own loyalty and are concerned that they might not belong in this organization. Members must know that all organizational members are accountable for their ethical behavior. Subordinates in an organization must have confidence and trust in organizational subordinates that management will hold all accountable for any malfeasance. The rules must be transparent to all, and checks and balances to deter unethical behavior should be intact.

Organizational members who are held accountable for their behavior perform best when positive performances are encouraged.

The most important thing leaders should know is that people will follow them. If such does not occur, then the leader is definitely not leading. Leaders need to learn what they can do differently if people are not following. Many books and courses are available to support the growth opportunities that exist to improve one's place as a leader. Never stop learning. When one thinks he knows everything, internal sirens go off that most often are ignored. Ego also gets in the way, and the success that could be gained is limited or null because of this.

Positive Influences

The key idea is for the leader to be endorsed by subordinates and authorized by superiors. Leaders should respect their followers by listening and collaborating. When leaders meet with any of their followers, they should do so with undivided attention; otherwise the impression will be instilled into the followers that they do not matter. At that point, the leader will most likely lose them as followers, and they will then become taskers.

Organizational leaders often misuse the talents of their employees, thus destroying the motivation to initiate any innovative ideas. This type of behavior also decreases the chances of adaptation to change as organizational resources tend to operate under the constraints of job security; thus, employees would remain unwilling to expose any undisclosed talents they may possess.

When members see unethical conduct in an organization, suspicion is created that changes their mindset from that point forward. Taking pride in the way one conducts business affairs creates self-satisfaction and self-worth; the results are immeasurable, providing a sense of accomplishment and enthusiasm in the organization.

An organizational leader's focus is on helping people perform more effectively and efficiently. One of the qualities an organizational leader must possess is the ability to communicate intuitively. The positive approach demonstrated when an organizational leader operates by an honest code of ethics promotes security in the hierarchical structure. When leadership sets an example, the members are loyal, and they will trust the leader to do the best thing for the organization. That is why it is important for the leader to be a positive example, and not confuse positive ethical practices with gray areas.

Leaders should carry out their roles with integrity. Followers want to give more than what is asked or required, and they will do so if leaders give them a reason to make a difference for the organization and themselves, with integrity.

Closing

Organizational culture, development, leadership—these are the foundation stones that any organization should rely on as very important aspects of success. Without a clear understanding of these facets, opportunities to achieve goals and successful outcomes may be missed. It is obvious that some organizations have processes in place on paper, but they implement something totally different, which causes an adverse effect in terms of what people do versus what they know to do. The information presented in this introduction and in the book series provides specific scenarios on how to make changes that will be very beneficial for guiding anyone to meet successful outcomes in the areas that are covered in these books.

Be empowered to make changes or know that you are on the right path as you seek to be the best in your organization. Read the book series, and contact Empower Us for other services to support your organization's goals. If you are new in management or integrating cultures, or if you want to confirm that the path you have taken is right for you, we offer the services and tools that will help you make the best decisions for your specific situation.

We hope you benefit from our information and give us feedback on other topics that could provide alignment and growth opportunities that lead to successful outcomes.

God bless,

Dr. Melanie.
Empower Us, Inc.
www.eulmc.com
contact@eulmc.com

Bibliography

Bolman, L. G., & Deal, T.E. (2008). *Reframing Organizations: Artistry, Choice, and Leadership.* San Francisco, CA: Jossey-Bass.

Hacker, S., & Roberts, T. (2003). *Transformational Leadership: Creating Organizations of Meaning.* Milwaukee, WI: ASQ Quality Press.

Magruder, M. (2013). *A Phenomenological Study: Information Technology Executive Leadership's Influence on Organizational Culture.* Ann Arbor, MI: ProQuest LLC.

www.ingramcontent.com/pod-product-compliance
Lightning Source LLC
Chambersburg PA
CBHW070341190526
45169CB00005B/1991